True and False Lights

D A L T O N   H I E T T

True and False Lights

PALMETTO
PUBLISHING
Charleston, SC
www.PalmettoPublishing.com

Paperback ISBN: 979-8-8229-5862-3
eBook ISBN: 979-8-8229-5863-0

TABLE OF CONTENTS

INTRODUCTION

The modern West is struggling for all to see. The problems are many. The solutions offered have all been tried before without success. For Christians in the West, attacks come from all angles. On one hand, the pressure of a secular society that mocks anything based on faith as laughable does all that it can to sow doubt in believers. On the other hand, competing religions, both new and old, are creeping in through the internet and immigration. Worst of all are the non-biblical beliefs that have successfully infiltrated the Church and led many Christians astray. Even the most devout Christian is going to feel the weight of attack in this environment. So far, the attacks we are suffering are not physical, thankfully. That does not discount the danger of psychological attacks directed against us. Churches are shutting down due to low membership.

Others are splitting both for understandable and questionable reasons. A side effect of failing to reject ideas that contradict clear Scripture is that the energy of the Church will die off, and members will stop coming. Even nonbelievers are feeling the pressure. The threat of war is constant. Government institutions that are supposed to be neutral are clearly not.

The secular world believes that civilization can exist without faith at its foundation. It believes that the "light of reason" is sufficient for civilization. Christianity is the bedrock the secular West was built on. Christianity provided its morals, work ethic, sense of fairness, and more. Christians should never feel insecure about this fact. Remember that the Bible says that Jesus is the light, not reason. The crippling problems assailing the West are the result of its quest to exchange the light of Christ for the light of reason. We who live in this time of transition from one light to another would be blind to miss the forming cracks. Large chunks of the American public question the results of both the 2016 and 2020 presidential elections. It is doubtful that the results of the 2024 election will be accepted either. Germany is deindustrializing, Ukraine is embroiled in a war with Russia, and Spain recently saw a bid for the independence of Catalonia. Many more examples of war, civil war, and instability can be found with a quick internet search. Adding to the political strife, inflation is impossible to ignore, and new job creation has slowed to a crawl. The statement that only women get pregnant is now a political one. Unscrupulous people spend great amounts of effort trying to entice even children to engage in filth. The young struggle to find their feet and start their adult lives. Affording even a small house is impossible for many of the younger generation. We are always sending enormous amounts of money to other countries and getting involved in conflicts that, under normal circumstances, would not be our concern. Countless foreigners fill our streets, many of whom we know full well to be dangerous individuals. Even those that aren't strain resources and take work from native citizens. As we transition from one light to another, the center of gravity changes. Without Christianity as its center of gravity, the Western world cannot continue to exist in the form it once held, let alone maintain things built on that previous foundation.

It is easy for a believer to feel discouraged in this quagmire. The situation is, however, not hopeless. Christianity came into existence in a pagan world. Before that, the ancient Israelites were surrounded by countless pagan idols and gods. Yet in that environment, they were able to more than succeed and flourish. Not only that, but Christendom, over time, managed to outshine the rest of the world both in spiritual and practical matters. Why should a Christian be afraid of the challenge now? When we encounter pagans, witches, Muslims, Hindus, and so on, we have no reason to feel fear. Any persecution we are enduring in the modern day has yet to even approach the persecution in antiquity. Christians of the past dealt with all of these, and all we need is to look back to them for advice. The Bible tells us to expect difficulty in life. Christians in the past suffered worse things than we have yet imagined, and yet they survived and carried the faith to us all the same.

We should not be discouraged, but we should not be idle either. Some Christians have taken this situation as an opportunity to sit idle and endlessly repeat, "It's in God's hands," or, "God is in control." As the saying goes, "God feeds the birds, but He doesn't put the food in their nest." Yes, God is in control, and one of His decisions was to delegate to believers' a degree of responsibility in this situation. God doesn't micromanage everything in our lives. He could, but He chose in the Garden of Eden to give us free will. He will hold us to account for our actions or inactions in this life. Do not be the wicked servant who chose not to fulfill his duties because he didn't think the master of the house would be back for a couple of days. If anything, hard times should spur us on to work harder, not breathe easy. The more difficult the moment, the greater both the responsibility and glory.

This book has two goals in mind. The first goal is to help Christians understand the various attacks on the Faith and how to combat them with confidence. The second goal is to point out theologically unsound attitudes rampant in the modern Church. In order to achieve both of these goals, a big-picture approach is needed. Some of these moving parts will feel irrelevant until viewed as part of a whole picture. Bear in mind that if we believe in an eternal and omnipotent God, we should seek to have an equally farsighted understanding of current events. We don't have an excuse for having short memories.

CHAPTER 1

Before we can set out on the work, first, we need to understand the job we are accepting. To do that, we first need to understand the environment we are going to be working in. The Western world treats science as the ultimate source of truth. A believing Christian should hold the Bible in that place, not science. Many, if not most, believers in the West make this mistake. When an atheist proclaims the superiority of reason over superstition, it is a common response of Christians of all denominations to use exclusively scientific arguments in support of the Bible. A perfect example of this is the Bill Nye vs. Ken Ham debate in 2014 at the Creation Museum.[1] In this debate, Bill Nye claimed that the laws of physics have not changed for a moment since ancient times. If the laws of physics have not changed, then the miracles in the Bible could not have occurred. Logically, this is flawed. No one was measuring or studying physics in that sense back then. Without direct observation or record, how can we know that the laws of physics remained the same at that time and place as they do now? This is an assumption, not a fact. Theology also holds that God created the natural laws of the universe and He can alter them at will. Bill Nye, in making this statement, showcased a profound ignorance of the theology he was criticizing. Ken Ham did get around to this argument. First, however, he named a scientist and an engineer who both believed in creationism. Later, he made mention of evidence that confirms the reality of a young Earth. By relying on the existence of Christian scientists, he made two mistakes. When he emphasized scientists, he downplayed the value of the philosophical and theological arguments for God. This action also reinforces the mindset that only scientific proof has value. In other words, this one act makes Bill Nye's materialistic argument sound stronger than it is. When he makes mention of evidence confirming a young Earth, he does the same thing yet again. Ken Ham clearly had good intentions and handled himself well in the interview. The issue here is a lack of proper perspective that many, if not most, Christians in the West share.

Why do we have such a tendency to play up the value of science in these areas and downplay theology and philosophy? What makes science superior in value to the other two?

An effective word for describing this attitude is *scientism*. Merriam-Webster's dictionary defines *scientism* as "an exaggerated trust in the efficacy of the methods of natural science applied to all areas of investigation."[2] In the past, philosophy and theology were considered vital to any complete education. Now, the average person likely cannot define what philosophy is. The average Christian only has a basic understanding of theology. Under the dominance of scientism, the scientific method and logic itself are conflated as being the same thing. Science was created as a tool of logic. Its purpose was to prevent incorrect ideas from being accepted as fact. It was not created to answer all questions, nor can it. The scientific method consists of asking a question, creating a theory to explain the question, and testing that theory. If that theory cannot be tested, science cannot be used to determine the truth of subject.

In spiritual matters, science rarely is able to test anything. The term then used to describe this situation is the word *unscientific*. Merriam-Webster's dictionary defines *unscientific* as "not based on or exhibiting scientific knowledge or scientific methodology."[3] This term is commonly used to insinuate that something is illogical or irrational. Based on the dictionary definition, however, it only means that something cannot be tested or measured by science. If science cannot test something, it cannot give an answer of any kind. Christian theology holds that God is not measurable, fully knowable, or in any way finite. God is unscientific, not because belief in Him is irrational, but because He is not measurable and therefore out of the reach of science.

To the materialistic atheist, anything unscientific is automatically not real. It is important to realize how irrational this attitude is. Even if we treat science as the ultimate source of truth, the furthest it can take us in the direction of atheism is to say that we can't prove God either way. This is not merely a mistake either. It is a product of willful ignorance. Most, if not all, atheists are emotionally invested in their disbelief. I have yet to make the acquaintance of any atheist that fails to lash out in anger when flaws in their logic are pointed out. Many of them claim that they would like to believe, but they need proof before they can. Provide any remotely convincing proof, and that same individual will grow red in the face. Perhaps there is an atheist that only lacks sufficient evidence for belief. If so, he or she is very lonely. Ironically, the accusation that religion is a crutch can be leveled at atheism. If no afterlife or higher power exists, then I can do whatever I want in this life without any consequence in the next. Perhaps the reality is that many atheists are afraid that the believer is correct and the nonbeliever is wrong. That would make for an interesting situation—both sides afraid of each other and doubting their own beliefs. It is important to remember people are driven by emotional concerns, not rational ones. Human beings have a strong tendency to choose what to believe and justify it later. Detaching yourself from your feelings and following the logic is not an easy task. No scientist can ever be fully impartial. They will always be influenced by their own beliefs and opinions. An atheist in a scientific field will feel the desire to use science to disprove anything spiritual. A Christian scientist will feel the need to use science to prove the Bible.

It is understandable for a materialist atheist to believe that science is the highest source of truth. Christians have no such excuse. We have another source of truth. Spiritual revelation in the past was enough for Christians. If it was written in the Word of God, that was enough. We didn't need proof because we were told directly from the source of truth. Now, for some reason, Christians need science to back up the claims of Scripture. In some Christians, there is a sense of desperation when confronted with a scientific criticism of the Bible. That fear is a symptom of having dual loyalties. The Bible and science cannot both be the ultimate source of truth. It may have been a long time ago, but we have a record of words spoken to human beings by the creator of the universe. If we are confident in our faith, we can trust in the Scriptures as their own source of truth. If these are the words of the God of creation, His words can be trusted more than science. Science is, after all, only the product of human ability. Have we been so wowed by technology that we would rather trust in man than God? Technology can't provide us purpose in life. It can't provide us answers about the spiritual world. It can't preserve us after death. These fancy gadgets we now have are nifty, but in the face of eternity, they are irrelevant.

Another forgotten point of interest is that, without philosophy and theology, there would be no science. Philosophy is the act of asking questions and working through those questions with logic. Much of philosophy is devoted to the proper use of logic in and of itself. Those involved with the creation of the scientific method were called philosophers in their day. Many of the same people were preachers, priests, and monks. Christianity has long held the idea that God created a logical universe. Studying that universe was an act of worship and a path to growing closer to God. Christian theology and Greek philosophy are the bedrock on which science was built. It was the religious desire to understand the world in an environment that valued logic that provided even the possibility of science. Without them, science itself is a lesser version of what it could be. One must ask how long it can endure without both of those two pillars it was built on.

If science is a hammer, theology is a screwdriver. Each exists to answer a different set of questions. When the atheist tries to apply science to the existence of God, he is hammering screws. Instead of falling into scientism, the believer should lean on philosophy and theology when trying to wrap their mind around God. Don't hammer screws. By the same token, don't screw nails. The Bible exists to provide us answers about God, the spiritual world, and our place in relation to eternity. It does make mention of physical processes and holds many historical records. Neither history nor science is the primary focus, and for that reason, the

Bible only gives us limited aid in understanding those fields. Archaeology is an ever-changing field, and it is logically possible that differences in archaeology are flaws in archaeology rather than Scripture. That is a matter of faith, not logic, however. Remember also that Scripture holds many possible interpretations, and we don't always know which interpretation is correct. The Bible doesn't make those things clear, because they are not the questions it is focused on answering. Be just as careful in applying the Bible to scientific subjects as you are applying science to spiritual subjects. Theology and science are, again, tools intended for different purposes.

In embracing scientism, the Western world is not only making a logical error. By doing this, the very foundation of Western civilization is in peril. Christianity and its teachings are the roots that feed the plant. Philosophy is the leaves collecting sunlight. By uprooting all spiritual belief, the West is cutting itself off from its very source of life.

CHAPTER 2

The materialistic mindset has a viciousness to it. For those that hold this viewpoint, reason and clear thinking are enough to run a civilization. In their minds, civilization is, in fact, improved by the absence not only of God, but culture, memory of the past, and loyalty to anything beyond individual needs. For them, history exists as only an ugly, unfortunate occurrence. Nations are inconvenient and even harmful things. The presence of religion or God are only suffocating restrictions on individualistic desires. The existence of different cultures is only a source of conflict. Better that all of these things just stopped existing. Not only do they believe these things need to go, but they also find the very ideas these things represent as repulsive. The attitude is summed up well in a quote from the noted atheist and physicist, Steven Hawking:

"One can't prove that God doesn't exist, but science makes God unnecessary."[4]

There is, of course, a fundamental flaw in this view of the world. When people ascribe to materialism and reason in this fashion, they assume that the structures a civilization is built on will keep working without God, nations, culture, shared history, or even solid identity. Perhaps the idea that such things are required for the continuation of civilization has never occurred to them. Perhaps their desire to remake the world in their utopian vision is so strong that they are willing to risk it. Not only is there a practical issue with this vision, but there is also a philosophical issue. Without the presence of a higher power, how do you define good or better? There is a desire to bring forth what they see as a utopian vision, but why is this vision a good thing to bring forth? They might defend their point of view by telling you that the past is painful, and it would be better to forget. They might tell you that religion causes wars and unnecessary suffering. The nation as an entity and the idea of borders are outdated. Even if we pretend such things are true, why are they bad? It's as if the only form of morality they can put into words is to say that pain is bad. Or maybe, a more accurate explanation would be that pain is unpleasant.

For what they claim to be morally good amounts to what feels good to them, and what they claim to be morally bad amounts to what feels bad to them.

Not only does the materialistic viewpoint rely on flawed assumptions and a lack of self-reflection, but it also holds to a naive understanding of human nature. At the core of this mistaken belief system is the idea that human beings are generally good. If this is true, then all conflict and suffering is the result of misunderstanding. This leads to the idea that if we just understood each other better, war would be obsolete. If we could just understand each other, we could solve poverty and raise the standard of living for everybody. Crime wouldn't exist if people weren't poor and uneducated. The idea exists for these people not only that people are naturally good, but that society is what makes people bad. In that mindset, the wrongdoer is innocent of his crime, and society, as a whole, is guilty instead. Here, again, we encounter that lack of reflection on morality. The Bible tells us that human beings are generally sinful. Jesus tells us that the poor will always be with us. Common sense agrees with both of those statements. You don't need to teach a child how to steal toys from another child. Parents have to spend effort to teach them not to. If someone was holding you at gunpoint, telling him you have a family will only help the situation if the gunman cares about your family. If he does not, you are wasting your breath. People are very capable of being intentionally evil.

If we are to understand right and wrong properly, we need to realize that morality requires an arbitrator. If there is no higher power, right and wrong becomes a confused concept. This is best explained with a thought experiment. Most people will readily agree that killing random strangers is morally wrong. Odds are, they have never stopped to think on why it's wrong. Christianity states that human beings are made in God's image. To destroy God's image, there needs to be a compelling reason. If God is not considered in our reasoning, then why is killing anyone wrong? Sure, it might make us feel bad when someone we care about dies, but what logical reason is there to say their murder was morally wrong? Christianity has an answer. Secular thought does not.

According to German philosopher Friedrich Nietzsche in his book *Beyond Good and Evil*, two forms of morality exist: master morality and slave morality. In Nietzsche's understanding, master morality exists when someone becomes his own source of morality, while slave morality is a system of rules weak people use to control strong people.[5] In this description, Christianity is a made-up set of rules designed to control people of strong will. In his mind, people of strong will have the right to do what they wish by virtue of their own strength of will. What Nietzsche is describing, in reality, is a world in which people are the arbitrators of their own morality. The strong can do what they like. Everyone deserves only what they can take for themselves. The only difference is whether something is gained through one's own strength or through trickery. Emotionally, most people reject this idea out of hand. But if no higher power exists, who is to say Nietzsche is wrong? If man is the master of his own destiny, the only morality that exists is the morality he places on something.

Many atheists in the modern day embrace this view and go out of their way to say that choosing our own morality in a universe that has none is actually beautiful. In reality, it is a shallow and depressing idea. People have an innate need to be a part of something greater than themselves. This form of morality will always leave people with only a sense of emptiness and lacking. Human achievement is impressive but still irrelevant when compared to the scale of the universe. The absence of anything of higher existence than ourselves does not elevate humanity. What it does do is describe a universe lesser in scope and grandeur than that imagined in any era prior to the Enlightenment. Let us not forget all the horrible mistakes made by human beings either. From the oldest records ever written to yesterday's news, human beings have continued unceasingly to commit atrocities upon each other. If we are the pinnacle of existence, the universe is a bland, bleak, and empty place. If alien life exists, this equation is unchanged. What reason do we have to believe that any alien civilization would be any less pitiful than our own?

Considering both the practical and emotional issues present in a Nietzschean view of morality, let's try another tact. Returning to a previous idea, most people agree that killing random strangers is bad. Why can that agreement not be enough? On its face, the argument sounds almost convincing. To put it simply, the problem is that people change their minds often. Imagine if everyone in a town signed a document stating that killing strangers is wrong. At first, this sounds fairly solid. But what if, over time, the townspeople began to think things over, and now, a majority agree that killing ugly people is OK? They then draft a new document with this exception to the rule, and two-thirds of the town sign it. The town council can denounce the new document and arrest anyone who attempts to kill an ugly person, but that is a practical matter, not a moral one. What is the source of morality here? Is it the signed piece of paper? If so, what makes the original document morally superior to the new one? Morality by majority opinion is, by its nature, always changing. In a majority-opinion basis of morality, the charismatic person can dictate morality to be whatever he or she wishes. A good speaker who has a way with words can easily convince a crowd to agree with his idea. In the West, we have been told from childhood that democracy is a good thing, but we, at our own peril, forget that the will of the people is not necessarily good. The majority of the crowd wanted to crucify Christ. The majority of the crowd, on several occasions, wanted to stone Paul. The majority of people in the 1700s believed slavery was good, and now, the majority believe it is wrong. The majority of Germans in the 1940s believed that placing Jews in concentration camps was a moral good. For democracy as a system to function properly, it must draw its sense of morality from a source other than itself.

Another problem is that in both of those two situations, self-interest is encouraged. In the Nietzschean view, those of strong will decide right and wrong for themselves. In the case of majority morality, all you have to do is convince 51 percent of the crowd. In either case, there arises the opportunity to choose the thing that benefits you as right and that which does not benefit you as wrong. For those in positions of authority, Nietzschean and democratic morality will inevitably only be seen as tools of power and control. This, in the end, goes to prove that morality based on reason is not morality at all. True morality must come from a source higher than and separate from those it judges. No system of government or ideology, not even democracy, can long exist without stable moral foundations at its base. Yes, ideology is also insufficient. New information always arises to challenge old ideas, and if ideology is a source of morality, it will be little more stationary than public opinion. The Nietzschean view comes out a little better in terms of stability, but still not overtly well. At least the Nietzschean view was honest about seeing grabbing power as an intentional goal. Ironically, rule by dictatorship can be more stable than democratic rule, as things will remain largely unchanged so long as the dictator lives. Of course, any moral system he devises will only last as long as he does. Everyone around him will inevitably have different interpretations on things, no matter how loyal they are to their idolized leader. As soon as he is gone, these different interpretations will come into conflict. This is all assuming that this strong-willed dictator is highly capable and has devoted supporters. Without these, he will be only a puppet of other forces and hardly in a position to decide morality even for himself, let alone others.

Not only does secular materialism undermine morality and stability in the West, but it also drains society of the energy necessary to maintain itself. In a Christian conception of morality, men have a duty to protect those weaker than themselves. The wealthy have a duty to provide charity for the poor. In fact, all people have a duty to help meet the needs of others around them. All of these are moral duties directly commanded to a Christian by God through the Scriptures. Similar things are true of other religions as well. If God does not exist, then what moral obligation does anyone have to do any of these things? Fewer and fewer men will feel the need to protect their nation or neighbors. The wealthy will no longer feel the need to give to the poor. Neighbors will lose any concern for each other's well-being. If politicians can make a fortune by skimming off the top for themselves, why should they not? Why should a policeman risk his life to save a random citizen from armed attackers? Why not lie and get someone you're angry at thrown in prison when given the chance? Many workers will cease to care about the quality of their work. The energy that gives a civilization life will begin to ebb away. A nation in this state will become sluggish, confused, and increasingly unable to hold itself together. Without the presence of a higher power in the minds of the people, no common goals or drive will exist. The result is a society that no longer cares about its own well-being and the inevitable slide into stagnation that follows.

CHAPTER 3

Science does not disprove the existence of spiritual things. Nor can it provide a sufficient foundation to build a civilization on. It is then no wonder that people still seek to find morality in spiritual sources even in a high-tech world. This does create another issue, however. Most spiritual experiences occur only in the mind of the experiencer. Compounding the issue, we cannot test the miracles of the past, as those would not have left much, if any, evidence as to whether they happened or not after so much time. Few religions claim that a miracle can be performed on command, which would be a requirement to test the existence of miracles. Several denominations of Christianity have, for example, held the doctrine of cessation (the idea that all miracles stopped after the age of the apostles) for quite some time. Without anything to test, science cannot answer the question of spiritual experience at all. Even the most hardened atheist or well-studied scientist will admit when pressed that science does not disprove spiritual realities. That is not enough on its own to believe, at least for me. Everything spiritual that can be tested, as short a list as that is, should be tested. Science will likely not be able to provide a solid answer to any spiritual question, but that is fine. Science should not be expected to fill our lives, especially considering it was never designed to do so. That does not mean that whatever hints or details it gives us should be ignored. We should embrace the testing of spiritual things when possible.

Scientific testing has produced at least some evidence supporting the possibility of spiritual realities. The CIA performed experiments on what they called "psychic abilities" in the 1970s as a part of Project Star Gate. Even readers who have heard of the project may not realize that many of the test subjects were practitioners of the occult.[6] Another important point is that the project's official report is that those abilities were real. Officially, the project shut down because those abilities were not consistent enough to be useful to the government. Considering how such projects go, who knows if they actually shut down? A common practice with secret programs that get exposed is to merely change the name of the project, claim it was shut down, and keep going without any significant changes. Adding to this possibility, government officials dabbling in the occult and accidentally contacting demonic forces might explain the behavior of some politicians that surround us in the current day. Perhaps the satanic panic in the 1980s was justified after all.

On to less unnerving prospects. There is tangential evidence that also supports the existence of spiritual forces. A good example is the experiment carried out by several organizations, which PEAR (Princeton Engineering Anomalies Research) is the most widely known for in this particular field. In these experiments, a device that randomly creates numbers (a random number generator) is placed in an empty building, and its numbers are recorded. After enough data has been recorded to understand the normal pattern of numbers the machine creates on its own, a group of people is brought into the building, sometimes for a planned activity like a church service or yoga session. The machine is shielded from vibration, static, electromagnetism, sound, temperature, and the air in the room by its special housing. Sometimes the participants are told about the machine, and other times, they are not. In both cases, the very presence of people in the building participating in a shared activity has been shown to change the pattern of numbers created by the machine.[7] Human thought, according to these findings, is able to affect the world without any physical contact. More than one organization has carried out its own studies, and seemingly, the results are repeatable. Maybe a physical explanation for this discrepancy will be discovered, or maybe human thought really is the cause of the change. Christianity will be true either way, but it is nice to have circumstantial evidence support the existence of spiritual things.

There is a new promising theory of human consciousness called the quantum consciousness theory.[8] This theory is an extension of quantum brain theory and a new discovery in how the brain works. According to quantum brain theory, the brain is a quantum computer. This is an extension of the discovery that parts of the brain are affected by quantum phenomena. This is important because of a feature of quantum physics discovered in the double-slit experiment. The results are quite complicated and difficult to explain adequately. The simple version is that photons are fired through two slits in a thin piece of metal at another thin piece of metal behind the first. The bizarre part is that the particles behaved differently depending on whether they were watched or not watched. When the experiment was not watched, the pattern of photons hitting the metal in the back was scattershot and acted like a wave. When the experiment was watched, the photons acted in a much more uniform fashion, like bullets moving in a straight line, as you would expect of particles. This is important because the presence of someone watching affects the ability of the quantum computer to work. You can't have the quantum structures involved in computing misbehave the moment you look away and behave when you look back. For a computer to work, certain things need to remain constant. So, if the brain is a quantum computer, who is watching? This observer (the person watching) must be watching at all times the quantum computer running for it to work, and people don't pass out when left in a room alone. Either that, or the computer needs to be shielded from all observers at all times, and people don't pass out when you look at them. Since the theory was proposed in the 1960s by physicist Hiroomi Umezawa, quantum brain theory has been largely disregarded. In the last few years, however, it has been discovered that certain brain structures (microtubules) are affected by quantum phenomena. If we consider the results of random number generator experiments with this new discovery, we arrive at an interesting possibility. The two men that first proposed the idea (Roger Penrose and anesthesiologist Stuart Hameroff) are of the opinion that this quantum consciousness is the result of only physical processes. An alternate interpretation of quantum consciousness theory is the idea that the soul is the observer that allows the quantum brain to work. This part of science is new, and maybe the theory will be proven false. Maybe it will not. Quantum computing itself is still getting off the ground, and we still have a lot to learn in this area. Not being either a physicist or a computer engineer, I can only speculate. Christianity can remain true either way, but it would be very interesting if it turned out to be true. The resurrection described in the Bible involves a new resurrected body given to believers. The corrupted flesh (the body we are born with) is replaced by new flesh (a new body not tainted by original sin) as opposed to existing as a spirit only. Quantum brain theory only gives one possible explanation that could match what Scripture says about the subject.

Assuming this interpretation of quantum consciousness theory is true, then it is logically true that spiritual experiences are as real as physical matter. If the random number generator study results are true, we have proof that conscious beings can, at least in small ways, affect the physical world through thought alone. It is logically possible that the soul and body combined through these brain structures create a human consciousness. A spiritual being, whether benevolent or malevolent, could also use the same structures to communicate. This experience would only be felt by the recipient, and the only evidence we would have would be the experience itself.

As interesting as the above facts are, none of them provide more than circumstantial evidence in favor of spiritual forces or beings. These discoveries are also very new and hotly contested. Science can only tell us a little about these spiritual experiences beyond some small details. We should test spiritual experiences when possible, but it is also important to avoid overly relying on science. Thankfully, science is not the only tool of logic available to us. Philosophy existed before science, and it has plenty to tell us.

In philosophy exists the idea of the prime mover or unmoved mover, found in Aristotle's two books, *Physics* and *Metaphysics*. According to Aristotle's logic, all things in nature are explainable by the chain of cause and effect. At some point, far back enough in history, there must have been a first cause to start the chain reaction. Natural forces only react to a prior cause. But if natural forces only react to something that acted on them first, then how did this first cause happen? Aristotle explains this conundrum by pointing out that living beings do not require external forces in order to act. They can choose to act. Therefore, the first cause is set into motion by a living being. This doesn't quite get us to God. In order to be the prime mover, this being only needs to be conscious, not necessarily intelligent. The need for a prime mover at the origin of the universe does not prove the existence of God, but it does open the door to the possibility of God. There are many philosophical arguments in favor of the God of the Bible, but they are too complicated to detail in this book.

At this point, we have a logical argument for the truth of the Christian faith that takes into account both scientific discoveries and logical exercises in philosophy. Many lifetimes could be spent pursuing study in both of these areas, let alone trying to connect the two. For now, we have a solid enough foundation to focus instead on making the case for Christianity above other religions.

CHAPTER 4

In the absence of miracles to test, the only way we can reliably judge the truth of these spiritual encounters is by examining the accounts of those that have them. The Bible, after all, does call us to test every spirit. The most obvious issue to navigate here is the fact that these spiritual experiences vary widely and contradict each other at many points. Our first bit of theology to cover here is the idea of lying spirits:

> *Therefore look! The Lord has put a lying spirit*
> *in the mouth of all these prophets of yours, and*
> *the Lord has declared disaster against you.*
> —1 Kings 22:23 (KJV)

Many varied religions now exist in the West that did not decades ago. Immigration and the internet have exposed the Western world to belief systems it never previously had to deal with. These religions are growing in number as well. Witchcraft is making a comeback. We have pagans again. We have Muslims, Hindus, Buddhists, and a smattering of cults. The rationalistic attacks on Christianity have hardly bothered to attack other spiritual viewpoints. The organization Queers For Palestine is a good example of this phenomenon. That organization is filled with homosexuals who support radical Muslims who are known to kill homosexuals on principle. All contradictions between these extremely different views on life are ignored so long as they continue to act as battering rams against Christianity. Refuting scientism is not enough. Christians must be able to refute these other religions just as we must be able to refute the atheists. Before we can test the spirits, these other religions need to be categorized. A chart explaining them is down below:

Category	Definition	Examples
Polytheistic	Many Gods	Greek mythology, Norse mythology
Pantheistic	All things are parts of God; individuality is an illusion.	Buddhism
Monotheistic	One God	Christianity, Islam, Judaism
Agnosticism	The supernatural world is unknowable beyond surface details.	Perineal philosophy
Animism	All things, including trees and rocks, have their own spirit and agency.	Various Native American beliefs

Looks like a mess, doesn't it? The best and most accurate way to navigate this mess is to compare all religions from the point of view of all other religions in order to see who has the best explanation, explaining not only its own positions but also the others. The scope of such an endeavor is far too large for this work. Instead of a comprehensive comparison, all other religions will be examined from the perspective of Christianity.

To begin with this comparison, the first belief system under the microscope is polytheism. In polytheistic faiths, like Greek paganism, many gods exist. Each of these gods possesses certain jurisdiction over the physical world and the characteristics that go along with it. Storms happen at sea often because Poseidon, the god of the sea, has a quick temper, for example. The characteristics of these gods change often from region to region and over time. In the Norse pantheon, the gods don't affect nature in such a direct fashion and explain human behavior instead of natural phenomena, for example. Regardless of what pantheon of pagan gods are being examined, they are always portrayed as significantly more flawed and less powerful when compared to the Christian God. It has long been Christian teaching that pagan gods are actually demons seeking worship and attention that should rightfully go to God. Demons, including the devil, are never described in the Bible as being anywhere near as powerful as God. If these spirits are lying spirits and possess far less power and character than God, any religion they start will look quite similar to the various pagan pantheons. It would also explain various pagan practices, including human sacrifice, cannibalism, holy prostitutes, and so on. The pagan point of view does not possess nearly as strong a counterpoint. Most pagan beliefs have a single creator god, but he is commonly weak enough to be usurped by his offspring. Many religions, like Hinduism, have a tendency to treat Christ as only one extra god, in the lowercase.

This idea contradicts the claims Christ made during His ministry, however. How can Christ be on the same level as all the other gods when Christianity makes clear that there is only one God?

The pantheistic view holds that all things are God. The trees are part of God. The rocks are part of God. People are part of God. God is everything, and everything is a part of God. In this view, God is not the creator but, instead, creation itself. One example of this would be nature worship. Another would be Buddhism. According to Buddhism, all suffering is the result of existing apart from God. In other words, you suffer because you are an individual, and your identity has not been absorbed into the blob yet. Christianity sees God as the Creator above and separate from the creation. For many, the idea of losing your identity and sense of self as the only way to escape suffering sounds terrifying. The emotional reasons can be enough to reject pantheism. Here, however, the goal is to outline a logic-based refutation of pantheism from a Christian point of view. Buddhism was not founded from a direct revelation. Instead, the teachings of Siddhartha Gautama began Buddhism. Siddhartha Gautama was not a divine figure and instead was a mystical teacher whose practices were supposed to allow one to shed their identity bit by bit in exchange for a sense of peace. All pantheistic belief systems tend toward this situation. In the end, no higher being made contact, and the religion is the result of human teaching on how to contact that higher reality. In Christianity, the teachings are considered the Word of God given through a human as an intermediary. Christianity also claims that God, in the form of Jesus, came to Earth in physical form. If God has directly spoken, then why should the words of a human, no matter how wise, be accepted instead? This is more than just a feature of Buddhism. New age and other pantheistic belief systems all share the mystical teacher as the central figure. Nature worship may be the exception to this, but there, you still have the problem of worshipping creation instead of the Creator. Christianity has a more authoritative source than pantheism.

Animistic beliefs claim that all things, including rocks and trees, have their own animating spirit, and everything in nature has a personality. By animating spirit, it is believed by the animist that the rocks, rivers, and trees are conscious and can act on their own. Animism differs from pantheism by claiming each of these things is its own separate entity and that such an arrangement is normal instead of a problem to be solved. Animism sometimes involves the spirits of the ancestors and sometimes a higher power. This higher power often acts more like a force than a living being to make things more complicated. The reason for this very vague definition is that animism is highly varied and mutable. Whereas pagan gods are many and sometimes change from location to location, in animism, the spirits change names, roles, and even personalities constantly over much shorter distances. One could say the spirits involved are different based on geography. Even then, this inconsistency still bodes poorly for its authenticity. Animism also shares similarities with polytheism in that the various spirits are never described as being as powerful or independent of creation as the God of Christianity. These spirits are even seen to be subject to nature as much as human beings. There is a two-pronged approach here. One, while the presence of such spirits is untestable, the claims of agency are testable. I am unaware of any rock having displayed agency. I have neither experienced a rock talking nor heard of any study in which one moved on its own without a convincing scientific explanation. The second point is that the nature spirits described in animistic faiths from the Native Americans to various African tribes share characteristics with the gods described in polytheism. They are flawed and far less powerful than the God of Christianity. The view that they are, in fact, demonic beings can be applied here as well.

In the case of agnosticism, we have something that sounds a bit more rational and testable. Agnosticism is the belief that the supernatural exists, but we can never know for sure anything of real detail about it. One example of this would be perineal philosophy. Perineal philosophy states that all religions draw from the same source of truth. By studying the religions of the world, we can find patterns and discover truths. That sounds almost like testing the spirits in a Biblical sense. There is an important distinction, however. If we remember the doctrine of lying spirits, we must realize that not all of those religions are equal sources of truth. If spirits are capable of lying, then not all spiritual experiences can be considered trustworthy. Why, logically, would spirits be more prone to speaking the truth any more than human beings do? Perineal philosophy, for this reason, can be rejected by virtue of failing to critically evaluate its sources. Perineal philosophy is, of course, only one version of agnosticism. Some agnostics don't think the spiritual world matters, even if it does exist. Some agnostics look at the various religions and simply say, I don't know who is right. This attitude can lead to trying a little bit of everything—maybe a bit of Catholic mass followed by a visit to the mosque before attending a Buddhist meditation session with their guru. This buffet-sampler mentality is essentially a less intellectual version of perineal philosophy. All forms of agnosticism can be boiled down to the act of asking the wrong questions or failing to ask the right ones.

We have now marked out all categories outside of monotheism. Monotheism, of course, includes more than just Christianity. Islam and Judaism are the two largest competitors. Zoroastrianism is an old competitor. Sometimes animistic religions have a single creator spirit and might qualify as monotheistic. Each one of these three faiths will be tested as their own case. The first monotheistic faith here to be examined is Islam.

Islam is the second-largest religion in the world. When it comes to Islam, there is much to criticize from the Christian point of view. Mohammad had a large number of wives, of whom his favorite was Aisha, the one he married when she was six years old. That, however, is an emotional and moral argument. Here, the goal is a logical argument. Islam, according to Muslims, is a continuation of Christianity. If that is true, then the Quran should have no major contradictions with the Bible. To examine this claim, a comparison of verses from the Bible and the Quran is required.

> *You are of your father the devil, and the desires*
> *of your father, you want to do. He was a*
> *murderer from the beginning and does not stand*
> *in the truth, because there is no truth in him.*
> *When he speaks a lie, he speaks from his own*
> *resources, for he is a liar and the father of it.*
> —John 8:44 (NKJV)

> *And no wonder! For Satan himself*
> *transforms himself into an angel of light.*
> —2 Corinthians 11:14 (NKJV)

> *But even if we, or an angel from heaven,*
> *preach any other gospel to you than what we*
> *have preached to you, let him be [a]accursed.*
> —Galatians 1:8 (NKJV)

From these three verses, the Bible claims three things. The devil is a liar and the father of lies. The devil is known to present himself as an angel of God. Finally, the gospel taught by Paul and the twelve apostles is intended to be the only gospel. No new teaching after that time should be taken seriously. Even if an angel preaches a different gospel, it should not be trusted. Next, here are a couple of Muslim passages.

> *The Prophet (ﷺ) returned to Khadija while his heart was beating rapidly. She took him to Waraqa bin Naufal who was a Christian convert and used to read the Gospels in Arabic. Waraqa asked (the Prophet), "What do you see?" When he told him, Waraqa said, "That is the same angel whom Allah sent to (the Prophet) Moses. Should I live till you receive the Divine Message, I will support you strongly."*
> —Sahih al-Bukhari 3392

> *But they (the Jews) were deceptive, and Allah was deceptive, for Allah is the best of deceivers!*
> —S. 3:54; cf. 8:30

Mohammad did not originally know if the spirit that spoke to him was an angel or a demon. He was only convinced it was an angel when he received advice from his pagan wife's cousin. Also, the Quran was written after the time of Paul and the early Church, making this a new gospel that should not be trusted. Finally, Allah in the Quran is described as the best of deceivers. Mohammad questioned if the spirit that gave him the Quran was an angel or a demon. The devil disguises himself as an angel. Mohammad was given the Quran after the time the Bible says no new gospel should be trusted. Allah is the best of all deceivers. The devil is the father of lies. From the Christian perspective, the answer is clear. Mohammad was given the Quran by a demon, possibly the devil himself! There are also the satanic verses in which Mohammad prophesied only to be told by the spirit from the cave that those words did not come from Allah and he should retract them. So not only is the messenger Mohammad encountered questionable in character, but he also admitted to prophesying things that the messenger never told him.

The Muslim counterargument is that the Bible is corrupted, and this is why Mohammad had to be given the Quran directly. Allah can't have someone corrupting Scripture again, can he? But why then does the Quran seem to confuse the birth of Jesus and the birth of Moses, for example? Many accounts in the Quran also share a striking similarity to Christian folktales that circulated during the time of Mohammad. There are many other textual problems with Islam, but for now, the case is strong enough.

Zoroastrianism differs from Christianity, Islam, and Judaism by claiming God (*Ahura Mazda*) is not all-powerful. He relies on human action to sway the tides of good and evil in his direction. Ahura Mazda is not the creator either; that was someone else. It also makes the claim that the first human was a giant hermaphrodite. The only reason Zoroastrianism is worth examining here is the fact that many modern-day scholars claim that Christianity and Judaism copied many of their beliefs from Zoroastrianism when the Jews lived in exile from their homeland. This position is tenuous. The oldest Jewish religious texts date back to the fourth century BC. The oldest Zoroastrian texts date back to the thirteenth century AD. It is possible that Zoroastrianism began in the sixth century BC, but even then, archaeological evidence is not conclusive. If anything, it is more likely that Zoroastrianism copied those similar features from the Jewish faith.

Now on to Judaism. Many Christians will struggle with what I have to say here. Nothing here is said out of hate or malice. I am not seeking to hurt feelings or put anyone in danger. By the same token, many nonbelievers will take issue with a laxer treatment of

Judaism. As stated previously, the goal here is to follow the logic, not to slander anyone, or to be irrationally sympathetic. Bear with me here. Starting at the beginning, the death and resurrection of Christ is the core point of separation. The Jews reject Jesus as the prophesied messiah. Christians accept Jesus as the Messiah. Over the centuries, things have gotten more complicated than just this, however. As Christians have the New Testament, modern Jews have the Talmud. The Talmud is a collection of rabbinical writing after the time of Christ. Some things included in the Talmud are magic rituals to be used by practicing Jews. Most of the book is a list of rules a practicing Jew should live by. Those rules have more than a passing similarity with the rules Jesus criticized the Pharisees for promoting. Some Christians struggle to remember that the Jews are not Christians. Adding to this, the modern nation of Israel, like all governments, has been guilty of immoral behavior. This does not make them a boogeyman. They are human and possess all the pitfalls that come with that fact. All the same, their goals and the goals of Christians are not the same. Christians share more in common with the Jews than with Muslims, but there are significant differences that must be remembered.

Considering the amount of time that has passed since the resurrection, we have very limited options in testing the reality of Jesus's ministry. Most of what we have is the written records of the pseudo-historian Josephus and the four Gospels. In other words, all we have is the words of people who lived there at the time of the events. Some argue that the questionable statements made by Josephus about his own life, combined with the religious nature of the Gospels, are enough to discount the entire account of Jesus and His ministry. Josephus does seem to have a habit of exaggeration. If we did discount Josephus and the Gospels, however, we would be left with very little to work with at all when trying to understand that period of history in the holy land. The written extra-Biblical records do not solidly prove the events of the four Gospels, but they hardly disprove those events either. Though not as solid as we'd like, the historical record does strongly suggest that Jesus was a real historical figure. Archaeological evidence gives a great deal of context for the records of the Gospels, but it once again cannot determine with certainty how accurate or inaccurate the four Gospels are. A romantic take on this situation is that God arranged this intentionally so that belief would be through faith rather than fact. Whether the miracles occurred or not cannot be tested. We can compare the recorded life of Jesus in the Gospels to the prophecies in the Old Testament. This, however, still relies on the accuracy of the Scriptures. I believe the Bible in its entirety is inerrant, but that is a matter of faith. Here, the goal is logic, and logically, we don't have solid proof. Jesus claimed divinity by calling Himself the Son of God. Logically, He either believed what He was saying, or He did not. If He did not, He was a liar. If He believed what He was saying, He was either a madman or He was telling the truth. These are the three options.

Jesus was executed for His claims to divinity, and all indications are that He saw the danger coming. Despite seeing impending death, He carried on anyway. Eleven out of the twelve disciples were executed for spreading the word of Jesus as the Messiah. The evidence we have suggests sincere belief on the part of both Jesus and His close followers. Either He was a madman, or He was telling the truth. Once again, we arrive at the position of an untestable question. How can the sanity of Jesus be tested? Currently, it cannot be tested. If He returns, then we don't have to test His sanity, as clearly, He would be more than human. Logic, in this case, only provides a path granting the possibility of truth in the Christian position rather than a solid explanation. Perhaps someone else has an explanation I have not heard. Perhaps this won't be solved until the Messiah arrives, either for the first or the second time. Addressing the divinity of Christ is not testable by science, but it does hold logic. Logically, if God created the universe and human beings, why could He not inhabit a human body as a method of communicating with His creation?

CHAPTER 5

It is obvious to anyone that the Western world is missing something important. A thorough picture of that missing thing requires an explanation of the world that existed in the ages before. Modern depictions of past societies tend to paint them as backward, stupid, superstitious, and needlessly cruel, with terrible hygiene. A more honest investigation will find that people of the past were little different from us. We have just as many idiots and vicious criminals now as they did then. They were just as intelligent as, if not more than, anyone living today. Nor were their lives unending misery compared to ours. There is evidence that the medieval peasant had more time off than the average worker today. They had a much slower pace of life than we do. Most of all, they had a more spiritually fulfilling environment than people living today. The desire to act as if the past was more horrible than the present comes from a knee-jerk reaction to defend the idea of progress. The truth is, for every step of "progress" made in one area, mankind has taken a step backward in another.

When we use the words *progress* and *backward*, we are assuming that history is moving toward a set destination. What destination are we talking about though? Why is technological advancement a sign that we are getting closer to that destination? Shockingly little thought is ever given to where we are going. Progress in the Enlightenment sense is mankind improving itself endlessly as history marches on with some vague paradise at the end when humanity has fixed all its own problems. The Biblical view of history has more than one interpretation, but none match this story. Even in the postmillennial interpretation (the belief that Christ will return after the world is sufficiently Christianized), humanity is improved not by its own efforts, but by the presence of the Church and the Holy Spirit. Most interpretations of Scripture declare that history is a downward slide into more depravity until the world ends. This idea of progress in the Enlightenment sense does not fit Scripture or history.

We now know that many of the things people believed in the past are factually incorrect. The mistake made by modern people is to think that everything they believed was wrong. The Greek philosopher Eratosthenes accurately measured the size of the Earth by measuring the shadows of several wells in two separate cities. Considering how much difficulty many of us have learning the Pythagorean theorem in school, it is safe to say that the Greek philosopher Pythagoras must have been rather intelligent to prove that formula. It took more than a thousand years after the fall of Rome to rediscover how to make concrete that sets underwater. Think about people you meet on a daily basis and ask yourself, Can they do this sort of thing? In terms of intelligence, we are by no means better than people of the past. Anyone who reads medieval philosophy will quickly be disabused of the notion that people of that era were stupid. As much as we might laugh at superstitions of the past, we are not free of superstition today either. How many people walk around a ladder instead of under it? How many people talk to their old car or lawnmower and beg it to start? How many people look and the sky and say, "Please don't rain today," as if the sky hears them? Human beings are superstitious by nature, and by laughing at past superstitions, we only make ourselves hypocrites.

We are told that people of the past were cruel and brutish. Some of them were. Many people of the current day still are. Human nature is the same now as it was then. We like to think we are better than those that came before us. For example, arranged marriages must have been barbaric. How awful that you might have no say in who you married. Compare the number of successful marriages of today to arranged marriages of the past. How many people today have been divorced and remarried? Divorce back then was rare. That modern voice wants now to say that it was morally wrong to force the woman into that relationship, even if it did work better. Arranged marriages did not always ignore the woman's wishes. If she voiced strong enough concerns and had family members with good character, the marriage would still be called off. She might not have a say in who she wanted to marry, but she usually did have a say in who she did not want to marry. Nor did the man always have a say in who he married either. Sometimes, he would get pressured by his family to marry a specific woman in place of his sweetheart. The biggest difference is that marriage was seen as a practical concern instead of a wish fulfillment. A good marriage was one that resulted in well-adjusted children. You didn't marry just because you felt attracted to that person. If the man couldn't hold a job, he wasn't a good match. If the woman was constantly drunk, she would make a poor mother. Hormones blind the young to these realities. By focusing on love as the reason for marriage, the modern world has forgotten the well-being of children in favor of fulfilling the fantasies of adults. Perhaps they were wiser than we are in such matters. Many of the old ways of doing things are like this. They are abhorrent to the modern mind, but if you think it through, it becomes clear that there was a good reason to do it that way.

A heavy cost of living in the modern world is the loss of moral consideration in our daily activities. In the mind of the prescientific world, the spiritual and physical interacted daily in almost every facet. Escaping the spiritual world's influence was impossible. It was believed that your appearance revealed your moral health. Ugly thoughts were believed to make a person ugly. By the same token, good thoughts made a person look good. To the modern mind, such a thing sounds ludicrous. Compare pictures of politicians before their careers and late in their careers. Do the same with non-politicians. You may find that this belief is at least partly correct. An alchemist attempting to create the philosopher's stone only for the gold it would produce would never succeed. It was believed that the heart condition of the practitioner was just as important as the technique. Another cost is the reality that our lives pass by like a streak of lightning compared to theirs. We are slaves to the future, never able to stop for a moment's rest without risking disaster just around the corner. Miss one paycheck, and you might lose the house. The hamster wheel must keep turning. Ironically, we have less say in our own schedules. The local lord didn't care how you lived your life or how many hours you worked. So long as you paid your taxes, and notified him when you traveled, he was satisfied. Now, most of us work with constant supervision in which every minute of our regimented workday is monitored and judged. They also had the privilege of living in a world that provided more room for spiritual nourishment than the one we live in. The spiritual world was not in doubt. To them, logic and revelation were two equally important tools for understanding the world.

In many ways, people of the past were more capable than people now. How many people can use a map in the age of GPS? How does handwriting before and after the keyboard compare? Relying on calculators can't have made people better at doing math in their heads. Before the ability to look something up on the internet, you had to remember what you read. The combined effect of modern advancements has made us, in many ways, less than those that came before. No one truly knows the full extent of damage done to the modern mind by all the gadgets we are so proud of. With those gadgets, we can do more than people of the past. But what did they cost us as people? So many of our modern inventions are so new that we can't yet have a clue what the side effects might be. Yet without any hesitation, most embrace them completely. Mocking the ways and people of the past is an exercise in arrogance and ignorance by modern people. Our knowledge has increased, but our wisdom seems to be fading by the same measure.

Of course, it wasn't all rose-colored. Famine and hunger were much more common. Diseases that are now easily cured were a death sentence. They had far less access to entertainment. Yes, in some ways, our lives are better. Solving those problems did not come free, however. Our lives speed on at a breakneck pace. We are less physically and mentally fit than our predecessors. Technology is abused as much as it is used properly. That smartphone can do many wondrous things. But it is also used to track your location and listen in on every word said in private. Nuclear technology can power homes or make them disappear. As an example of how horrible and abusive science can get, both the Japanese Unit 731 during World War II and the United States Public Health Service in the Cold War injected people with diseases, either against their will or without their knowledge. In both cases, it was done to study the effects as part of a scientific study. Modern technology is amazing and clearly a benefit to people. The cost of obtaining it was, by extension, high. We have traded strong families, spiritual fulfillment, peace, and moral advancement for electronics, cars, the internet, and air conditioning. The modern act of mocking people of the past is a coping mechanism. The deal we collectively made hasn't paid off like we thought it would. It cost us far more than we realized it would. At the same time, we can't bear to part with the product we purchased. We would have suffered if we didn't take the deal just as much as we suffer now. The truth is that life will never be easy. It's just a different kind of hard. That better life people of all ages have looked for never comes true, regardless of what we do. The reality of this situation is hard to reconcile with. Instead of looking for a better life in this world, perhaps it would be better to look forward to a better life in the next. The Bible outright tells believers to expect to suffer in this world. Let's untangle our understanding and put our focus on the prize.

CHAPTER 6

The lack of wisdom, morals, and motivation caused by scientism is causing incalculable damage to the Western world and Church. The influx of new ideas and belief systems is only compounding the problem. Even if the West is unable to recover from its current problems, all Christians still have a duty to ensure the Church outlives the civilization it exists in. To do that, confusion in the Church must be banished. Banishing that confusion requires a bit of history.

We now return to Aristotle. Before the scientific method, the teachings of philosophers like Aristotle were the primary method for understanding the physical world. The Aristotelian method is a complicated system for explaining how to use logic. The philosopher practicing Aristotle's method sought to answer questions with logic alone. The biggest difference between Aristotle's method and the scientific method is that Aristotle's method did not include experimentation. It was focused on the proper use of logic above all else. The idea that the Earth rotated around the sun was considered. Most philosophers from Aristotle on rejected the idea because we should be able to feel the Earth moving if that was the case. Without modern tools or advanced telescopes, this idea made more sense than the reverse. Scientism often sees Aristotle as an early figure on the way to progress and science. This view is biased and colored by Enlightenment thinking. More on that later.

Unlike the scientistic worldview we are accustomed to, the Aristotelian method didn't stop at explaining the physical world but also attempted to explain how the spiritual and physical worlds overlap. The existence and actions of the gods were an equal study to that of mathematics. When Rome conquered Greece, this method did not die off. Greek thought conquered Roman minds just as Rome conquered Greek lands. In Rome, Greek became the language of the educated. After the collapse of Rome, the medieval world continued to embrace Aristotle's method. The result of Christianity and the Aristotelian method blending was a deep exploration of both physical and spiritual affairs.

The medieval understanding of the world is difficult for modern people to wrap their heads around. In the minds of medieval people, there was no separation between the physical and spiritual. Both worked together as a unified whole. The sun revolved around the Earth because the Earth was the crown jewel of God's creation. The Bible doesn't say that, but the philosophers worked out that it did. Disease was caused by the corrupting prescience of evil or the imbalance of the four humors. If you got the balance of elements perfect in an object, you would have gold. If you traveled far enough from civilization, monsters might still be found. Angels and demons could at times physically cross into our world. Words had magic power, and the older the language, the more powerful the words. People of low character could raise the spirits of the dead and interrogate them. Items touched by a saint had miraculous powers. The movement of the planets dictated events and people's behavior. A doctor's confidence decided the outcome of treatment as much as the treatment itself.

In a way, that world had a more complete explanation of things than we have. It was willing to explore realities that we now rarely even think about. It was at the very least a more exciting way of looking at things.

This understanding of the world was being chipped away at for decades before it was finally shattered by a man named Galileo Galilei. The fact that the Earth rotates around the sun nowadays isn't a big deal. In those days, Galileo, with this one discovery, put into question everything intelligent people had spent nearly two thousand years working out. Textbooks often claim Galileo was in danger because he disagreed with Church dogma, but it's more complicated than that. Galileo's discovery didn't contradict the Bible. It contradicted the agreed-upon conclusion held by almost all educated people from the 300s BC up to the 1500s AD. If the most educated minds of that huge stretch of history had all gotten something as basic as the Earth revolving around the sun wrong, then what else could they have gotten wrong? Is anything mankind has ever believed true? It is hard to imagine a modern-day comparison of a single discovery able to cause such mayhem. A close one may be to one day realize that your entire life has been spent living in a computer simulation. Galileo's discovery was a cataclysmic event for the people that lived to see it.

In the aftermath of the intellectual carnage Galileo caused, a new crop of thinkers came onto the scene. One of these thinkers was Rene Descartes. Rene

Descartes was a French philosopher and mathematician who lived from 1596 to 1650. In a twist that doesn't often reach the textbooks, Descartes arrived at his greatest discovery through a series of mystical dreams. In this series of three dreams, Descartes encountered a hooded figure he called the spirit of truth that showed him various psychedelic visions. Descartes's interpretation of those visions led him to his method of skepticism. This method became a major inspiration of the scientific method. Take a moment to savor the fact that the scientific method was created not by rational thinkers fighting superstition. Instead, at least one of its creators made his contribution by sharing mystical visions he had one night with his peers.

Descartes made many other discoveries, largely in math, but this new model of skepticism was his crowning achievement. It was believed at the time that with this method of skepticism, the shocking discovery made by Galileo could never be repeated. There now existed a method to prevent false theories from being accepted ever again. As he got older, Descartes grew embarrassed of the dreams that gave him his breakthrough. Despite this, he never became a materialist. Instead of abandoning belief in God, he sought to create a new metaphysical understanding of the universe that included the spiritual world. He wanted to mend the shattered world left in the wake of Galileo's discovery. He just wanted the basis for this understanding to come from reason instead of a mystical vision. Ultimately, Descartes was not able to repair that rift. Having not studied his writings in detail, I can only speculate whether his theories were incorrect or simply failed to gain acceptance by his peers.

A few decades after Descartes came another thinker that had just as much impact. Baruch Spinoza was a Sephardic (Portuguese) Jew who, for various reasons, lived among Spanish Christians after going into exile from the Jewish ghetto in Portugal. When it comes to the writings of Spinoza, it isn't completely clear what he believed. Like the ink blot on the paper, what you get from Spinoza can depend on what you want to see. The model of the universe described by Spinoza provided the basis for more than one line of thought. One strain his work led to was deism. Deism is the idea that after God created the universe, He left, never to return. Everything from then on was in human hands. Another interpretation of Spinoza's work is a pantheistic worldview: a world in which all things are a part of God and that God is creation itself rather than the creator. The third interpretation is the idea that there is no higher power, and only physical matter and the laws of nature exist. It must mean something that the third interpretation eventually became dominant among Enlightenment thinkers. Spinoza himself doesn't seem to have offered any clarification as to which interpretation he intended. None of these three interpretations include a God active in the affairs of the world. Considering the crowd that made the most use of his work later, it is possible that they wanted a world without any unexplainable elements in it and, for that reason, desired a world without God.

Between the works of Descartes and Spinoza, the Enlightenment movement was sparked. Taking from Descartes his framework of skepticism and the worldview provided by Spinoza, the Enlightenment as a movement made two promises. First, the Enlightenment promised that this new light of reason would banish the darkness of ignorance and superstition, leading to a better version of humanity. Second, this new light of reason would improve the tools available to mankind, and with those tools, everyone's lives would be improved. The price required to obtain these two promises was the tearing down of traditional institutions and ways of thought. How much tearing down differed between members of the movement. Some, like the admirers of Descartes, wanted to reform these old systems by only replacing features when new discoveries disproved the old beliefs. Some wanted to tear down everything that was not completely new. They wanted to pull up by the root all thought and tradition that came before Galileo as if everything from the old world was poisoned. In their minds, only a radical clean slate was enough. On the more radical side of the Enlightenment was the spark for both materialistic atheism and scientism.

The first fruits of the Enlightenment came into existence in the form of the American and French Revolutions. The American Revolution was brought about by the more conservative strain of Enlightenment thought. The American founders wanted to reform traditional institutions, not rip them out by the roots. The American Revolution was also extremely religious in nature, by virtue of the first Great Awakening that had ended only a decade before. One interpretation is that in early America, there was an attempt to synthesize a new, complete worldview that combined the spiritual and the physical into one umbrella explanation like the one that existed before Galileo. Many different views existed at the time, and there was in no way a unified understanding of what the revolutionaries wanted. As we are told, the Continental Congress involved many fistfights and shouting matches. Despite the lack of consensus, both Enlightenment philosophy and religious fervor played a significant role in the early founding of America.

When the French Revolution began, Americans supported it at first. After rivers of blood began to flow in Paris, that support began to slow. The French Revolution was much more determined to overturn the past, and there was no religious revival present to temper the radical elements. It was inspired by the more liberal figures of the Enlightenment, such as Jean-Jacques Rousseau. It was also led by radical figures embodied by Maximilien Robespierre. It didn't take long after the execution of the royal family before the different factions of the revolution began to turn on each other. The American Congress had fistfights. The French Revolution had mass executions. By the time it was over, tens of thousands were dead; the royal family was dead, and, in its place, Napoleon was crowned emperor. From monarchy to republic to empire, France ended up at nearly the place it started, with only a large pile of bodies for the effort. Reason alone was not enough to form the paradise that the radical strain of the Enlightenment wanted to create.

That spark of materialism still had not entirely taken over. It had, however, been fanned into an actual fire.

The two major figures to arrive next were Immanuel Kant and Georg Wilhelm Friedrich Hegel. Kant was a Christian, but what kind of Christian is difficult to pin down. On one hand, he still claimed to believe in the existence of the soul and of God. On the other hand, he spent most of his time attacking theological ideas that had been in existence for centuries. His criticism of the Catholic Church was very harsh. One of Hegel's main points of emphasis was the need to verify everything. We can't know anything is true unless it is verified. Hegel's view on Christianity was revealed in his writing *Life of Jesus*. In this book, Hegel speaks of Jesus as more of a wise teacher than a divine figure. He treats the miracles in the Bible as metaphors instead of actual events. The resurrection of Christ is not even mentioned. He claimed to believe in the existence of God, but he could hardly be called a Christian. Between the two of these philosophers, we see what was happening in the minds of influential people in the wake of the French Revolution that both Hegel and Kant were alive to witness. Doubt was being put on not only the Aristotelian method but Christianity as well. Those living a century after the Enlightenment were not dissuaded by the bloodshed and failure of the French Revolution. Instead of taking a step back from the Enlightenment, they were uprooting larger and larger sections of the previous world.

Another blow to the Enlightenment was the American Civil War, which occurred after the lives of Hegel and Kant but only eighty years after the revolution. Preventing the less Enlightenment-influenced South from abandoning the project required another tidal wave of bloodshed. Over half a million Southerners and Northerners died in the conflict, totaling almost one out of every fifty people living in the country. Even in success, the North fundamentally altered the fabric of American society. As it is often said, "Before the Civil War, the United States are, and after the Civil War, the United States is." It can be argued America as an Enlightenment project never fully recovered from the Civil War. The light of reason failed to prevent a bloody war. The side of reason, in fact, initiated that bloody war to reign in the more religious and traditional dissidents. The Civil War is, of course, more complicated than this summary, but differences in adherence to the Enlightenment between the North and the South are inarguable.

Decades after the end of the American Civil War came two new influential thinkers. One almost all readers will recognize: Karl Marx. The other was a name mentioned in an earlier chapter, Friedrich Nietzsche. The ideas of these two philosophers set the world stage for World War II. Unlike the previous thinkers, between Nietzsche and Marx, there was no room left for discussions on spiritual things. The focus now was only on the physical and what science was able to achieve. We had now arrived at full-blown materialism. Most know that Marx inspired Communism. Fascism was the result of changes some former communists wanted to make to Communism. Fascism would not exist without Marx. Much less well remembered is that Nietzsche inspired National Socialism, the ideology of Nazi Germany. The push for democracies and republics from the 1700s to the 1800s was largely the result of the Enlightenment. Fascism, National Socialism, and Communism from the late 1800s through the 1900s were also the result of Enlightenment thought. Marx and Nietzsche were strongly influenced by Kant and Hegel. Kant and Hegel were strongly influenced by the Enlightenment.

World War II, from this point of view, was the battle between different children of the Enlightenment.

Democracies like America and Britain in this period only differed in the sense that their forms of government were older products of the Enlightenment. Communism, Fascism, and National Socialism were the younger siblings. By the end of this conflict, Fascism and National Socialism were dead. Along with those belief systems, fifty to sixty million people were dead.

During the Cold War, only democracy and Communism were left. The West was the older child, and the East was the younger sibling. America portrayed itself as individualistic, possessing greater freedom and liberty than any other system. Sometimes, Christian values would be held up, but freedom and liberty were front stage. The USSR prided itself on the improved version of mankind it intended to produce and the level of equality that man would surely have when Communism had reached its end goals. Given the differences between the two, it is easy to forget that both systems share many core features. Both are the offspring of the Enlightenment, and neither are willing to challenge its ideas. The conflict was between their differing methods of carrying those ideas out to completion. By this point, the influence of Enlightenment thought had spread to Asia in the form of Communism. South America had strains of both communism and democracy. The contest between the United States and the USSR spared only Antarctica's involvement. The revolutions, famines, civil wars, and proxy wars during the Cold War claimed more lives than World War II. Many records from the period were destroyed, so it is difficult to know the full death count. In China, it was anywhere from 15 to 43 million dead. In the Soviet Union, it was somewhere from 20 to 80 million. In Cambodia, another 1.5 to 3 million. The twenty years of war in Vietnam saw between 791,000 and 1,141,000 dead Vietnamese. The American dead is officially numbered at 58,300. These examples (the list is not complete) give a total of over 37 million dead on the low end and more than 127 million on the high end. My personal guess is that the middle number of 82 million is probably close to the real number.

The democratic governments, while not guilty of causing mass death, were not innocent either. Secret documents from the Cold War that have been made public reveal a pattern of manipulation and experiments designed to control the minds of their people just as deeply as the Communist governments. The US government drugged US and Canadian citizens without their knowledge or consent in Project MK Ultra.[9] Many other examples exist of this behavior and would take a great deal of time to cover. The only difference in behavior between the democracies during the Cold War and the Soviet Union is that the Communists used terror while the democracies used manipulation and secrecy.

With the collapse of the Soviet Union, only democracy was left as a major player. The growing truth now, however, is that even in a world with no competition, democracy cannot fulfill both promises of the Enlightenment. If it could have, Communism, Fascism, and National Socialism would likely never have existed, considering democracy came first. The French Revolution was an abject failure. The American democratic project has had some success, but that success falls short of the promised goal. Fascism and National Socialism are dead. Communism has made a comeback, but it is past its prime all the same. Doubt about the potential of the American project at the time of this writing is spreading. Democracy is the oldest and most successful child, but even it has failed to live up to expectations. Just as it started to look like the Enlightenment might succeed in its promises post-Soviet Union, 9/11 shattered the illusion again. Just in case anyone still clings to the hope that the United States will succeed in bringing about world unity through democracy, the rise of China and the Russian invasion of Ukraine should put the final nail in that coffin.

The promise that science and reason would improve the available tools and material comforts of the average person was fulfilled by the Enlightenment. The invention of painkillers and indoor air conditioning most certainly has made life easier. The promise to improve humanity itself, however, has been an abject failure. The two World Wars were concrete proof that reason and science don't make war obsolete. They only make it more savage and efficient at killing. People have not grown wiser or more intelligent, nor has knowledge of medicine and anatomy made the average person more fit and healthy. Obesity in the West is now rampant when it used to be rare. If anything, people of the current day are inferior in ability to those that came before them. The dominant understanding of the world in the current age is dull, empty, and mechanical compared to the pre-Galileo understanding.

The Enlightenment failed one of its two promises, and yet the price was paid upfront without a clause for a refund. The shattered pieces have still not been put back together—neither has a new world successfully been forged. At the time of this writing, people still talk endlessly about Communism, Fascism, and democracy. The reality is that all three of these systems are either dying or dead. They all came into existence for the purpose of fulfilling the Enlightenment, and they all failed. Governments that use these names increasingly don't live according to the label they take up.

Those that have heard of the World Economic Forum may now realize that the vision espoused there is a new attempt to fulfill the Enlightenment. Most that hear the new vision find that it sounds more terrifying than hopeful. It describes a world in which the average person has less say in their own life, has lower quality food, has less access to heating or cooling, has a reduced right to travel, and so on. In return? Free food and the internet appear to be about it. Not even good free food is offered. Nothing spiritual is offered at all. Nothing but guaranteed physical survival is on the table. If this new world they intend to build is too depressing, they are including suicide pods you can use to take the easy way out if this new life is too hard. The dying embers of the Enlightenment can now only propose meeting basic material needs. To quote a Scripture verse: "But He answered and said, 'It is written, "Man shall not live by bread alone, but by every word that proceeds from the mouth of God."'" (Matthew 4:4 [NKJV])

The world, as of now, is starved for spiritual sustenance. The light of reason is anemic, giving light but not warmth. By now, Enlightenment values do not have the ability to fulfill the promises they make, even for those that believed the exchange would have been a good one to begin with. The world needs to move on to something else. A new way of viewing the future is needed. This view needs to account for all the scientific discoveries made in the last few centuries. More importantly than that, however, it must make room and provide guidance for spiritual matters. Man cannot live on bread alone. Mankind cannot live on material prosperity alone. The spirit needs to be nourished just as badly as the body.

CHAPTER 7

In the Old Testament, the ancient Israelites repeated a cycle over and over. Every time things became difficult, they would turn back to God, and by the end of that generation, things had made a comeback. After a few generations of living in prosperity, obedience to God began to fade once again. Before long, things got difficult again. Only after things became unbearable did they return to God and begin the cycle again.

According to Neema Parvini, this cycle happens to all civilizations. In his book *The Prophets of Doom*, he examines the writings of various authors on the subject across time and geography. All these writers, with different words, according to Parvini, describe the same pattern. At the beginning of all civilizations is a spark of religious zeal that gives it life. The successes of that civilization over time provide the method of their own destruction. After the wave of religious zeal fades, prosperity continues for a while. Intellectualism starts to ask questions and undermine what is left of that religious zeal. Without that zeal, the civilization begins to fall apart. The men at the beginning are of a rugged nature. The men at the end are more academic in nature.[10] His conclusions are more complicated than that, but these are the takeaways useful here. Faith is the source of life for civilization.

It is good to reflect that this scientistic attitude is not unheard of in history. According to Paravini, the "barbarism of reflection" is part of the cycle that all civilizations go through. Science in the West has produced technology far beyond the imaginings of other civilizations. This can lure us into thinking of science as something newer and more special than it is. If all civilizations far enough into development suffer intellectuals tearing at the foundations, we are, at most, experiencing the same thing to a higher degree. And yet at the end of every civilization, a spark of religious zeal gives birth to a new one. In a way, the pressure to doubt spiritual realities is artificial. It is less a thought-out logical conclusion and more of an attitude that comes and goes with the ages. For now, we must deal with it, but it will pass.

Not all civilizations are equal. Some have risen far higher than others. If faith is the source of life for civilization, then maybe some faiths are of a higher quality than others. Western civilization has risen higher than any other before it. Many of those successes are directly traceable to Christianity. Circumstantial as it is, this is evidence in favor of Christianity as being the truth. The worrisome part of this business is that, according to *The Prophets of Doom*, the cycle never appears to stop or reverse when in motion. Perhaps ancient Israel was unique in this regard. The Old Testament makes it clear that Israel had a unique position with God. Western nations, unfortunately, do not have that unique position. It should be clear that we are much closer to the end of the cycle than the beginning at this point. It is likely already too late to rescue the Western world from its death. At best, we may be able to slow the cycle and keep things going a little while longer. We must be ready for the collapse. Considering the West has risen higher than any previous civilization, that fall might be harder than that of previous civilizations as well. Truly difficult times may be on the way.

For many Christians, the imminent return of Christ is a signal to take a break. Matthew 18 and the parable of the wicked servant beg to differ. We are told that, upon the return of Christ, those servants who are working on behalf of their master will be rewarded. Those that are found ignoring their duties will be dealt with harshly. The possibility remains that this is not the end of history. We may only be experiencing the natural cycle of civilization. In either case, troubling times are not a moment to sit back and wait. When this subject has come up it is common to hear the retort, "It's God's will." The people who repeat this phrase seem to believe that they no longer need to be concerned about the situation. We should not despair, but ignoring the situation is not what we ought to do either.

Is it God's will that people turn away from Him? Would He not prefer them to turn to Him instead? When God gave us free will, he gave us the ability to make choices. God's will occurs when we align our choices with His desires. God has a plan for when we obey and another plan for when we disobey. He would prefer that we obey. He is prepared for when we do not. If God tells us He is going to pass judgment on people, is He not telling us only that He knows they will not choose to repent?

Instead of acting as if the events in the world around us are not our concern, we need to be aware of our surroundings. We need to be aware that many wicked people are working hard to trick us into believing lies. If that does not work, they create enormous amounts of noise to confuse us. Our first duty is to weather these attacks. Do not be led like a dog on a chain. Do not serve the devil in ignorance. Many churches shut their doors during the COVID-19 lockdowns when the government politely asked. Even during the lockdowns, we were told that essential services would stay open. In what world is the Church not an essential service? During the lockdowns, many people found themselves in the hardest moments of their lives. Shuttering the churches at that moment was no service to them. It was an abdication of duty. The attitude in many churches is that politics should stay out, and yet the statement that only women can give birth is now a political statement. Should the Church now avoid biblical teachings on differences between men and women? No. Politics, for better or worse, have come for the Church, and ignoring them is not an option.

In the high likelihood that the West will collapse either in the near or faraway future, Christians should be involved more, not less. Most Christians are not in a position of authority. That does not mean they don't have responsibilities. We each are expected to do what little we can. If all you have to offer is a kind word to someone who is struggling, that can go a long way. If raising your children takes all the time and energy you have, make sure you are raising them with the Scriptures and protecting them from worldly lies. If you don't have children, you can still try to reach people you meet with the Gospel. Give the best effort you can and rely on God for the things outside of your ability. Whatever you do, don't pretend as if things in the world don't concern you. For those in a position of authority, you have a responsibility to use that position for Christ. What sense does it make to believe that God would be upset if you would use that position for His work? Would it not make more sense if God placed you in that position precisely to do His work? We are called to be in the world, not separate from it. Don't despair at the end of the West, but don't act as if you're exempt from it either.

CHAPTER 8

Now that we understand the world around us and how we got here, it is time to examine where we should be going. The Enlightenment, as a movement, became dedicated to tearing down old institutions and ways of thought. Christianity, having come before the Enlightenment, inevitably got caught in the crosshairs. Obvious attacks, including the efforts of materialist atheists' attempts to disprove Scripture, sadly have caused some to doubt. Far more dangerous are the ideas that come through the backdoor without our conscious minds noticing. Enlightenment ideas have sneaked into the Church and led it astray from God's design. Most of these unbiblical ideas that have successfully infiltrated the Church all boil down to the desire for equality. For many readers, this is likely difficult to accept. All of us in the West have been told from childhood that equality between people is inherently a good thing. It sounds like a good thing and can bring you warm, fuzzy feelings. The truth is that the Bible only promotes equality in one area:

> *There is neither Jew nor Greek, there is*
> *neither slave nor free, there is neither male nor*
> *female; for you are all one in Christ Jesus.*
> —Galatians 3:28 (NKJV)

Many will use Galatians 3:28 to say that God doesn't care about our differences. What it says is that the Gospel is for all people, regardless of station or walk of life. Being part of God's kingdom does not make those differences go away. When the Bible mentions all nations, tribes, peoples, and tongues being present in God's kingdom, it acknowledges that those nations, tribes, peoples, and tongues are different from each other. When the Declaration of Independence said that "All men are created equal," it was espousing an Enlightenment idea, not a Biblical one. God did not create men and women equal in strength, and neither did He give men the ability to give birth. Not only this, but God also did not make all men equally strong or all women equally beautiful. People are not equal in intelligence either. God also created differences in language when He scattered the builders of the Tower of Babel. God is the author of differences between human beings, and for that reason, we should be able to recognize and respect those differences instead of trying to prove otherwise. Perhaps He created more than just men and women in a complementarian fashion. Whether one believes in evolution or design, we should be able to recognize and respect that these differences exist for a reason. There is no good scientific or theological reason to believe that all people are equal in all respects, as the Enlightenment or egalitarianism would have us believe.

In the Middle Ages, the fusion of Greek philosophy and Scripture brought about the idea of the great chain of being. The idea is that all created beings are links in a single chain. God is at the top, with angels below Him, humanity below them, and animals below us. Some people are links higher up the chain than others as well. Being lower on the chain is not a sign of shame. Your value in the chain depends on how well you perform your role in the position you have. A diligent janitor is morally superior to a lazy president. Every link higher in the chain has more rewards, but more responsibility as well. Each person has their own role and place in relation to the whole. Your relationship to God is judged by how well you perform your own role, not by how many good deeds you do compared to your neighbor. The quest for equality can only result in pulling each link on the great chain apart and laying them on a table side by side. By doing this, animals are raised to the same value as human beings. By the same token, it lowers angels and God Himself down to the level of human beings. The quest for equality creates a desire to elevate the creation and reduce the Creator. Both situations are a detriment to humanity rather than a benefit. Without our place in the great chain to judge ourselves by, we instead can only compare ourselves to those around us. It is easy to end up feeling inadequate when comparing yourself to someone with twice the money, intelligence, work ethic, or looks as you do. Not only is God the author of difference, but equality is also itself, in many ways, less than desirable. Instead of questing for equality, Christians should be concerned with occupying their role and performing it well. To drive the point home, a few examples will be given here:

> *Husbands, love your wives, just as Christ also*
> *loved the church and gave Himself for her.*
> —Ephesians 5:26 (NKJV)

*Let a woman learn in silence with all
submission. And I do not permit a woman
to teach or to have authority over a man,
but to be in silence. For Adam was formed
first, then Eve. And Adam was not deceived,
but the woman being deceived, fell into
transgression. Nevertheless she will be saved
in childbearing if they continue in faith,
love, and holiness, with self-control.*
—1 Timothy 2:11–15 (NKJV)

In the biblical understanding, men and women are not equal but complementary. A man is expected to be willing to risk his life for his wife. We can deduce that he is also expected to risk his life for his children and maybe even strangers. A woman is not expected to risk her life for others. A woman is expected to be quiet and submissive. A man is not expected to be quiet or submissive. The Biblical view is that women should not have authority over men. Men should have authority over women. In the quest for equality, many in the churches have been led to mistakenly believe that having women in leadership positions is a good thing. It is rare that anyone would bat an eye at women in any position below pastor in an evangelical church of today. When the prophetess Deborah led the armies in the book of Judges, the Bible describes this event by leveling shame on the men of that time for failing to perform their responsibilities, not to praise Deborah for taking charge. The Bible says that men and women are of equal value but not ability. Men are stronger and better fit to lead. Women can make new people, and without them, there would be no next generation. Many nonbelievers turn away from the Church not because of its lack of equality for women but because of its weak men.

Far too many churches in the modern day have fallen into treating ideas of the Enlightenment as if they were Biblical commands. Some churches smuggle illegal immigrants into Western countries in the name of loving thy neighbor, despite the meaning of that passage, not because of it. Some have female pastors or priests. Democracy is held up as a universal good, even as the Bible describes heaven as a monarchy with God as king. The will of the people is likewise considered sacrosanct as opposed to the will of God. By adopting these worldly morals, Christians make idols for themselves.

CONCLUSION

The Western world is in desperate need of a new direction of travel. The Enlightenment promised to create something better than what existed before the discoveries of Galileo Galilei. After four hundred years of attempts, it should be clear by now that the Enlightenment was unable to fulfill that promise. Science and reason are incapable of improving human nature. In war, the power of science only makes killing more efficient and easier. In economics, reason cannot solve poverty. In academics, education cannot root out stupidity. In ethics, logic cannot provide for us the differences between right and wrong. Spiritual needs under the Enlightenment regime are discarded out of hand. People are now so starved for spiritual experience that dead religions and cults are rising from their graves. In short, the Enlightenment not only failed to fulfill its promises but what it promised wasn't enough in the first place. The West needs to move on to something else.

The old world cannot return. The new tools of technology don't easily meld with the way of life that existed before the Enlightenment. Even if they did, we now know much of that world's way of thinking was incorrect. There is no point in playing pretend or living according to something you know isn't true. Leaving behind those tools would also make the West vulnerable to other parts of the world that hold onto them. Enlightenment ideas instead need to be replaced with new ideas. The Enlightenment made promises in exchange for a price. In many ways, it was a knee-jerk reaction to the sense of loss people of the time felt at the passing of the Aristotelian worldview.

Before a new way of thinking can be pursued, the old things of the Enlightenment must be put aside. The will of the people is an Enlightenment idea. Placing focus on democracy or republicanism is an act of continuing Enlightenment thinking. As explained earlier, the worship of equality is also a core Enlightenment value. A radically new way of thinking needs to be found. This new system needs to learn from the mistakes made by both the Enlightenment and the Aristotelian worldviews. The primary failure of the Aristotelian system was a lack of ability to double-check conclusions. The Enlightenment failed to fully consider the needs of human beings to begin with and, worse, failed to fulfill its own promises. Perhaps a better approach would be to ask first, What do people need most in their lives? Not just in terms of physical needs but spiritual and mental ones as well. Only after that question is answered should we set out to determine how to provide those things. This new way should also be willing and able to double-check its own conclusions. Any such vision is far beyond the scope of this book.

Galileo was able to create an open wound in Aristotelian thought because the Aristotelian world was vulnerable to being wounded. The Enlightenment only allowed that wound to fester. To be successful, whatever comes next needs to provide for the spirit as much as it does the body. It must view science as a useful tool instead of the answer to everything. It needs to incorporate logic and faith together instead of treating the two as opposing forces. It needs to sift through the past and hold onto the truth that still exists there while at the same time embracing new discoveries. The current situation might last another couple of centuries. It might end tomorrow. Hopefully, the new way will arrive soon. Hopefully, it will succeed where its predecessors failed. Only time will tell. Until then, the best that anyone can do is to prepare for that new reality and survive the coming storm intact.

For Christians, the two important takeaways here are clear. We need to be able to recognize what ideas come from Scripture and what ideas come from the Enlightenment. Our morals need to come from the light of Christ, not the so-called "light of reason." The second takeaway is just as important. The Church as a whole needs to realize that politics and religion are not separable. They never have been separate, and they never will be. Political philosophy from the Renaissance dominates the minds of those in the Church now just as deeply as Aristotelian philosophy dominated the minds of churchgoers in the Middle Ages. Pursuing the goal of keeping politics out of the Church is a dereliction of duty, not a virtue. There is a great fear that letting politics in will corrupt the Church, and that is a valid fear. That is why it is vitally important that Christians be able to understand which moral arguments come from Scripture and which ones come from the world. Until the Church is able to unentangle itself from the Enlightenment, no real success can be expected.

NOTES

1. "Bill Nye Debates Ken Ham—HD (Official) 2014," interview by Ken Ham, 2014, https://www.youtube.com/watch?v=z6kgvhG3AkI

2. *Merriam-Webster*, s.v. "scientism," accessed April 10, 2024, https://www.merriam-bster.com/dictionary/scientism.

3. *Merriam-Webster*, s.v. "unscientific," accessed April 19, 2024, https://www.merriam-webster.com/dictionary/unscientific.

4. ABC News, "Stephen Hawking: 'Science Makes God Unnecessary,'" September 10, 2010, https://abcnews.go.com/GMA/stephen-hawking-science-makes-god-unnecessary/story?id=11571150.

5. Nietzsche, Friedrich. *Beyond Good and Evil: Classic Illustrated Edition* (Heritage Illustrated Publishing 2014) Kindle Edition, Chapter 9.

6. Beaumont, Dr. Roger A., "Cnth?: On the Strategic Potential of ESP." CIA, 1998, https://www.cia.gov/readingroom/docs/CIA-RDP96-00788R001700360003-3.pdf.

7. Nelson, R. D., R. G. Jahn, B. J. Dunne, Y. H. Dobyns, and G. J. Bradish, "FieldREG II: consciousness field effects: replications and explorations." *Journal of Scientific Exploration* (1998), 425–454.

8. "(U) Project MK-ULTRA," October 2011, https://www.cia.gov/readingroom/docs/project%20mk-ultra%5B15545700%5D.pdf

9. Paravini, Neema, *Prophets of Doom* (Exeter UK: Imprint Academic, 2023), 191–192.

ABOUT THE AUTHOR

Dalton Hiett, author of *True and False Lights*, has been on a spiritual and intellectual journey since childhood. Raised in a deeply religious environment, he later encountered secular perspectives which challenged his faith. This experience, during his college years, sparked a profound interest in asking and exploring questions others often overlook. Hiett sees himself as a conduit rather than a creator, and his work is the result of extensive reading, listening and research. Through his book, he invites readers to join him in exploring these thought-provoking questions and hopefully, start a dialogue on their own.

www.ingramcontent.com/pod-product-compliance
Lightning Source LLC
Chambersburg PA
CBHW071926120726
48001CB00005B/1884